Synthesis Lectures on Ocean Systems Engineering

Series Editor

Nikolas Xiros, University of New Orleans, New Orleans, LA, USA

The series publishes short books on state-of-the-art research and applications in related and interdependent areas of design, construction, maintenance and operation of marine vessels and structures as well as ocean and oceanic engineering.

Fidaa Karkori

Ship Vibration 4

Slamming Loads and Strength Assessment

 Springer

Fidaa Karkori
Southampton, UK

ISSN 2692-4420   ISSN 2692-4471   (electronic)
Synthesis Lectures on Ocean Systems Engineering
ISBN 978-3-031-74765-6   ISBN 978-3-031-74766-3   (eBook)
https://doi.org/10.1007/978-3-031-74766-3

This Springer imprint is published by the registered company Springer Nature Switzerland AG
The registered company address is: Gewerbestrasse 11, 6330 Cham, Switzerland

If disposing of this product, please recycle the paper.

Preface

This book has been written as a short guide which describes a slamming strength assessment procedure for vessels. Typically, there are three types of slamming loads on commercial vessels: bottom slamming, bow flare slamming, and stern slamming depending upon the hull geometry at the bow and stern. Upon verifying Class mandated compliance with the requirements outlined within this book, optional notations may be assigned to a vessel upon request by the vessel owner.

A fundamental requirement of this book is that the scantlings of the hull structure are expected to be in accordance with the criteria as specified in the Class Rules pertaining the classification, design and construction of marine vessels.

> The results of this book are not to be used in reducing the basic scantlings obtained from the application of the Class Rules. That is, if the results from this book indicate the need to increase any basic scantlings, the increased scantlings should be implemented above the Class Rule requirements.

Southampton, UK   Fidaa Karkori

The original version of the book has been revised. A correction to this book can be found at https://doi.org/10.1007/978-3-031-74766-3_10

Contents

Abbreviations and Acronyms

A/D	Analogue to Digital
AC	Alternating Current
BEM	Boundary Element Method
C_F	Conversion Factor
CG	Centre of Gravity
COMF	Comfort (Class Notation)
COMF+	Comfort Plus (Class Notation)
cpm	Cycles Per Minute
DC	Direct Current
FE	Finite Element
FEA	Finite Element Analysis
FFT	Fast Fourier Transform
FPU	Floating Production Units
HAB	Habitation (Class Notation)
HAB+	Habitation Plus (Class Notation)
HP	Horsepower
Hz	Hertz
ISO	International Standards Organisation
kN	Kilo Newton
kPa	Kilo Pascal
kW	Kilowatt
m/s	Metre(s) Per Second
mm	Millimetres
MRA	Maximum Repetitive Amplitude
MSDV	Motion Sickness Dose Value
PRU	Power Related Unbalance

Rpm	Revolutions Per Minute
SNAME	Society of Naval Architects and Marine Engineers (US)
TBC	Top Bottom Centre
TDC	Top Dead Centre

List of Figures

List of Tables

Slamming Phenomenon

1.1 General

In rough seas, the vessel's bow and stern may occasionally emerge from a wave and re-enter the wave with a heavy impact or slam as the hull structure comes in contact with the water. A vessel with such excessive motions is subject to very rapidly developed hydro-dynamic loads. The vessel experiences impulse loads with high-pressure peaks during the impact between the vessel's hull and water. Of interest are the impact loads such as bow flare slamming, bottom slamming, stern slamming, green water and bow impact loads.

These impact loads are of a transient nature and can cause severe structural damages. Although the loads are widely varying in their characteristics - magnitude, rise time, duration, etc. - all involve the impact at high relative velocity between the free surface of nearly incompressible seawater and the hull structure.

The transient impact loads can be highly non-linear and may be strongly affected by the dynamic response of the hull structure. Thus, magnitudes of impact loads/pressures are important for structural design purposes. The following items are considered for the determination of the impact pressures:

- Intensity and duration of impact pressures,
- Spatial distribution and time duration of impact pressures,
- Equivalent static slamming pressures for direct strength assessment and scantling determination, and
- Bow and stern geometry.

© The Author(s), under exclusive license to Springer Nature Switzerland AG 2025
F. Karkori, *Ship Vibration 4*, Synthesis Lectures on Ocean Systems Engineering,
https://doi.org/10.1007/978-3-031-74766-3_1

For vessels possessing significant bow flare, an impact occurs on the side plating of the bow as it is rapidly immersed into the water. This action results in a large fluid pressure covering a comparatively large impact area. There are two types of bow flare impact pressures: non-impulsive pressure and impulsive pressure.

The magnitude of the non-impulsive type pressure is directly related to the bow submergence while the magnitude of the impulsive type pressure rises rapidly at contact and decays exponentially in time.

High slamming pressure can be experienced on stern and bottom hull structures as a vessel emerges and re-enters the water. In this book, bottom slamming is considered to occur in the bottom structure in the region of the flat bottom forward of $0.25L$ (L is the Rule length, as defined in the Classification Vessel Rules) measured from the forward perpendicular (FP). Due to the stern and bottom slamming, two noticeable effects may occur. First, there may be localised structural effects in the areas of the stern and bottom that experience large slamming pressure. This may result in set-up plating and buckled internal frames, floors and bulkheads. The second effect of the slamming is the vibratory response of the entire hull, or so-called whipping, which is not within the scope of this book.

This book can be applied to ocean-going vessels including oil carriers, bulk carriers, container carriers, and gas carriers. The basic principles and requirements in the Classification Vessel Rules are to be reflected in the hull structure. The following slamming loads are covered in this book:

- Bottom slamming for all vessel types when the heavy weather ballast draft forward is less than $0.04L$,
- Bow flare slamming for container carriers, liquefied gas carriers, or other vessels possessing significant bow flare, and
- Stern slamming for vessels such as container carriers with significant overhanging sterns or liquefied gas carriers with relatively flat stern cross sections.

For vessels with the stern and/or stern geometry features that deviate significantly from the conventional arrangements, the scope of direct strength assessment against slamming loads is to be determined in consultation with Classification.

The design slamming pressures on the bottom, bow flare, and stern are defined as the most probable extreme loads on those areas during the 25 years of service life in the North Atlantic. A direct slamming strength assessment under this unrestricted service condition is required.

This book provides a rational, step-by-step, direct calculation method for the slamming strength assessment of the bottom, bow flare and stern structures.

Chapter 3 provides recommendations for the loading conditions, speeds, and headings. The locations to be analysed are explained in Chap. 4. Environmental conditions are described in Chap. 5, which is followed by the calculation of vessel motions in Chap. 6. Statistics of motion is explained in Chap. 7.

The calculation of design slamming pressure is explained in Chap. 8. Then, the strength assessment is covered in Chap. 9.

1.2   Optional Classification Notations

In recognition of full compliance with the strength criteria in Chap. 9, optional class notations may be granted. A summary of requirements and a brief description is given in Table 1.1.

Table 1.1 Indicative notations, description and requirements

Indicative notation	Description	Requirements
SLAM-B	Strengthened against Bottom and/or bow flare slamming.	The vessel needs to satisfy the Classification bottom slamming and/or bow flare slamming procedure and criteria in the bow area, depending on vessel type and hull form.
SLAM-S	Strengthened against stern slamming.	The vessel needs to satisfy the Classification stern slamming procedure and criteria in the stern area.

Nomenclature

2

2.1 General

The following symbols are used in this Guide. For some symbols, additional details are given where the symbols are used in this Guide. Also, some symbols, such as symbols for constants, are not listed in this chapter. Refer to the description with each formula for those not listed.

2.2 Nomenclature

A_c Effective shear sectional area of the support or of both supports for double-sided support, $(=A_{lc} + A_{ld})$, in cm^2 (in^2)

A_{lc} Shear connection area of lug plate $(=f_1\ell_c t_c)=$ f1ℓctc, in cm^2 (in^2)

A_{ld} Shear connection area excluding lug plate $(=\ell_d t_{tw})$, in cm^2 (in^2)

A_S Attached area of the flat bar stiffener, in cm^2 (in^2)

A_S Cross-sectional area of the spring rod, in cm^2 (in^2)

A_{shear} Effective shear area of a hull girder cross section, in cm^2 (in^2)

AP After Perpendicular

B Breadth of vessel, in m (ft)

B_{wl} The greatest moulded breadth measured amidships at the scantling draft, in m (ft)

C_{3D} Three-dimensional correction factor $(=0.83C_L)$ for slamming pressure

C_a Permissible bending stress coefficient

C_b Block coefficient

$\Delta_s/(1.025LB_{wl}ds)$ in metric tons and m

$35\Delta_s/(B_{wl}ds)$ in long tons and ft

© The Author(s), under exclusive license to Springer Nature Switzerland AG 2025
F. Karkori, *Ship Vibration 4*, Synthesis Lectures on Ocean Systems Engineering,
https://doi.org/10.1007/978-3-031-74766-3_2

C_d Plate capacity correction coefficient (=1.0)

C_l Location factor for bottom slamming pressure near FP

C_p Local pressure coefficient

C_s Dynamic load factor, as described in Chap. 8, Sect. 7

d_i Vertical distance from the still water surface to the location, in m (ft), if the location is above the water surface, then it becomes zero.

d_m Mean draft of vessel, in m (ft)

d_s Scantling draft of vessel, in m (ft)

E Modulus of elasticity of the material, may be taken as 2.06×10^7 N/cm^2 (2.1×10^6 kgf/cm^2, 30×10^6 lbf/in^2) for steel

f Spreading function for short-crested waves, as defined in Chap. 5, Sect. 3

f_1 Shear stiffness coefficient in slot connection check, or an adjusted yield point of material in shell plating check

f_2 An adjusted yield point of material in shell plating check

f_b Calculated ideal bending stresses in buckling check, in N/cm^2 (kgf/cm^2, lbf/in^2)

f_b An adjusted yield point of material in stiffener check

f_{bdg} Bending moment factor in stiffener check

f_c Collar load factor for slot connection check

f_{ci} Critical buckling stress with respect to uniaxial compression, bending or edge shear, separately, in N/cm^2 (kgf/cm^2, lbf/in^2)

f_{Ei} A buckling stress with respect to uniaxial compression, in N/cm^2 (kgf/cm^2, lbf/in^2)

f_y Minimum specified yield point of the material, in N/cm^2 (kgf/cm^2, lbf/in^2)

f_{Lb} Calculated uniform compressive stress in buckling check, in N/cm^2 (kgf/cm^2, lbf/in^2)

f_{LT} Calculated total in-plane shear stress in buckling check, in N/cm^2 (kgf/cm^2, lbf/in^2)

f_{yi} A yielding stress with respect to uniaxial compression, bending, or edge shear

f_u Ultimate tensile strength of material

FP Forward Perpendicular

g Gravitational acceleration, 9.81 m/s^2 (32.2 ft/s^2)

H Response Amplitude Operator (RAO) of a response

H_s Significant wave height, in m (ft)

l_v Moment of inertia amidships, in m^4 (ft^4)

K_i Buckling coefficient, as given in Chap. 9, Table 9.4

ℓ Length of stiffener between effective supports in buckling check, in cm (in.), or

             Unsupported span of the frame, in m (ft), or

             Spacing of transverse in slot connection check

ℓ_c	Length of connection between longitudinal stiffener and lug plate in slot connection check (refer to Chap. 9, Fig. 9.3), in cm (in.)
ℓ_d	Length of direct connection between longitudinal stiffener and transverse member in slot connection check (refer to Chap. 9, Fig. 9.3), in cm (in.)
ℓ_s	Length of spring rod in main supporting member check, in cm (in.)
ℓ_t	Cargo hold length between transverse bulkheads in main supporting member check, in cm (in.)
L	Scantling length, the distance measured on the waterline at the scantling draft from the fore side of the stem to the centreline of the rudder stock, in m (ft) For use in this book, L is not to be less than 96% and need not be greater than 97% of the extreme length on the waterline at the scantling draft. The forward end of L is to coincide with the fore side of the stem on the waterline on which L is measured. In ships without rudder stock (e.g., ships fitted with azimuth thrusters), L is to be taken equal to 97% of the extreme length on the waterline at the scantling draft. In ships with an unusual stern and bow arrangement the length, L, is to be specially considered.
LS	Longitudinal spacing
$Ltons$	Long tons in US units
m_n	n-Th spectral moment at a main heading angle
n	Number of spring rod elements in main supporting member check, or a constant for pressure–velocity relationship
n_s	A factor in stiffener check
P_s	Design slamming pressure, in N/cm^2 (kgf/cm^2, lbf/in^2)
$(\overline{P_s})_{max}$	Maximum of average slamming pressure over each location
P_1	Load transmitted through flat bar stiffener in slot connection check, in N (kgf, lbf)
P_2	Load transmitted through shear connection in slot connection check, in N (kgf, lbf)
P_r	Proportional linear elastic limit of the structure, may be taken as 0.6 for steel
s	Spacing of longitudinal or transverse frames in stiffener check, in mm (in.), or spacing of longitudinals/stiffeners in slot connection check, in cm (in.)
S_m	Strength reduction factor for various steel grades
S_y	Spectral density function of a response
S_ζ	Wave spectrum, in m^2-sec (ft^2-sec)
SM	Section modulus of stiffener
SM_{pl}	Section modulus of stiffener considering plasticity
t	Required net thickness of web plating in buckling check, in cm (in.), or total duration time in estimation of extreme response, in sec
t_c	Thickness of lug plate in slot connection check (refer to Chap. 9, Fig. 9.3), in cm (in.)
t_n	Net thickness of the plate in buckling check, in cm (in.)

t_{tw}	Thickness of transverse member in slot connection check (refer to Chap. 9, Fig. 9.3), in cm (in.)
T_2	Mean period of relative vertical velocity, in sec
T_p	Peak period of wave ($= 1.408T_z$), in sec
T_z	Zero crossing period of wave, in sec
v	Relative vertical velocity between water and hull, in m/s (ft/s)
V	Vessel speed
V_d	Design speed of the vessel, in knots
W	Width of the cut-out for an asymmetrical stiffener in slot connection check, measured from the cut-out side of the stiffener web (see Chap. 9, Fig. 9.1), in cm (in.)
α	Aspect ratio of the panel (longer edge/shorter edge) in shell plating check
α_b	Local body plan angle measured from the horizontal, in degrees
α_p	Correction factor for the panel aspect ratio
β	Wave heading, following sea is 0, and head sea is π, in rad
β_0	Main wave heading of short-crested wave, in rad
Δ	Vessel displacement, in tons (L tons)
Δ_i	Virtual displacement, including added mass of water, in tons (L tons)
Δ_s	Vessel displacement at the scantling draft, in tons (L tons)
μ	A constant used in 2-node vibratory frequency calculation
v	Poisson's ratio, may be taken as 0.3 for steel
σ_{fb}	Flat bar mean stress, in N/cm^2 (kgf/cm^2, lbf/in^2)
σ_r	Standard deviation of relative vertical motion, in m (ft)
σ_v	Standard deviation of relative vertical velocity, in m/s (ft/s)
τ_{dc}	Direct collar plate mean stress, in N/cm^2 (kgf/cm^2, lbf/in^2)
ω	Frequency of wave, in rad/sec
ω_1	2-Node hull girder vertical vibratory frequency, in rad/sec
ω_e	Encounter frequency of vessel motion, in rad/sec
ω_p	Peak frequency of wave $(=2\pi/T_p) = 2\pi/Tp$, in rad sec

Loading Conditions, Speeds and Headings

3

3.1 General

The dynamic slamming loads approach involves the selection of critical conditions for slamming load prediction. These conditions give due consideration to the vessel speeds, wave headings, and loading conditions.

3.2 Critical Loading Conditions for Slamming Load Prediction

In general, the critical loading conditions are to be identified based on the susceptibility of the hull structure to slamming pressure, giving due consideration to the fore and aft hull forms as well as the local drafts relative to the hull structure. Some of the characteristics depending on vessel type are:

- Oil carriers, bulk carriers and gas carriers are susceptible to bottom slamming due to shallow draft in ballast loading conditions,
- Container carriers are susceptible to bow flare slamming due to large flare angles, and
- Container carriers, gas carriers and passenger vessels are susceptible to stern slamming due to flat overhanging stern.

In this book, loading conditions are determined by slamming types as shown in the following.

F. Karkori, *Ship Vibration 4*, Synthesis Lectures on Ocean Systems Engineering, https://doi.org/10.1007/978-3-031-74766-3_3

3.2.1   Bottom Slamming

For the direct strength assessment against bottom slamming, a seagoing loading condition with a minimum draft forward is to be identified from the Loading Manual. Such a loading condition is typically associated with a ballast loading condition with a minimum heavy weather ballast draft forward.

If the minimum draft forward indicated on the shell expansion drawing is smaller than the minimum draft forward in the aforementioned sea-going loading condition in the Loading Manual, an additional loading condition with the smaller draft is to be developed solely for the purpose of direct strength assessment against bottom slamming.

3.2.2   Bow Flare Slamming

For the direct strength assessment against bow flare slamming, two loading conditions are to be selected: one is the sea-going loading condition corresponding to the scantling draft and the other is the sea-going ballast condition. If there is no ballast condition for the vessel type under consideration, for example, container carriers, a seagoing condition corresponding to a shallow draft with light cargo is to be selected from the Loading Manual.

3.2.3   Stern Slamming

For the direct strength assessment against stern slamming, two loading conditions are to be selected: one is the sea-going loading condition corresponding to the scantling draft and the other is the sea-going ballast condition.

3.3   Standard Speed Profile

In high seas, the vessel speed may be reduced voluntarily or involuntarily. For the slamming load prediction in this book, three standard speed profiles are to be applied based on the significant wave height and vessel length as shown in Tables 3.1, 3.2 and 3.3, where V_d is the design speed.

For vessels other than containerships with length exceeding 350 m (1,148 ft), the speed profile is to be included in the Loading Manual as guidance to the Master. This speed profile is for vessels operating in head seas subject to bow impact.

For stern slamming, lower speeds are known to be more critical than high speeds. Therefore, 0 and 5 knots are to be used for all wave heights.

Table 3.1 Standard speed profile for slamming load prediction for large vessels (L_{pp} > 320 m (1,050 ft))

Significant wave height H_s	Speed	
	Bottom/bow flare slamming	Stern slamming
$0 < Hs \leq 6.0m(19.7ft)$	100% V_d	0 and 5 knots
$6.0m(19.7ft) < Hs \leq 9.0m(29.5ft)$	75% V_d	
$9.0m(29.5ft) < Hs \leq 12.0m(39.4ft)$	50% V_d	
$12.0m(39.4ft) < Hs$	25% V_d	

Table 3.2 Standard speed profile for slamming load prediction for smaller vessels (220 m (722 ft) < L_{pp} ≤ 320 m (1,050 ft)

Significant wave height H_s	Speed	
	Bottom/bow flare slamming	Stern slamming
$0 < Hs \leq 4.0\,m(13.1\,ft)$	100% V_d	0 and 5 knots
$4.0m(13.1ft) < Hs \leq 7.0\,m(22.9\,ft)$	75% V_d	
$9.0m(29.5ft) < Hs \leq 12.0\,m(39.4\,ft)$	50% V_d	
$12.0\,m(39.4\,ft) < Hs$	Max (25% V_d, 5 knots)	

Table 3.3 Standard speed profile for slamming load prediction for smaller vessels L_{pp} ≤ 220 m (722 ft)

Significant wave height H_s	Speed	
	Bottom/bow flare slamming	Stern slamming
$0 < Hs \leq 3.0m(9.8ft)$	100% V_d	0 and 5 knots
$3.0m(9.8ft) < Hs \leq 6.0m(19.7ft)$	75% V_d	
$6.0m(19.7ft) < Hs \leq 9.0m(25.5ft)$	50% V_d	
$9.0m(29.5ft) < Hs$	Max (25% V_d, 5 knots)	

3.4 Wave Heading

It is known that the slamming in the bow area is more severe in head sea conditions, while that in the stern area is more critical in following sea conditions. In this book, a heading range from beam sea to head sea is to be considered for bow flare and bottom slamming, and a heading range from following sea to beam sea is to be considered for stern slamming.

Extent of Hull Structure to Be Evaluated

4

4.1 General

The extent of the hull structure to be evaluated depends on the types of slamming and is to be determined before starting the actual calculations. A minimum of at least eight frame stations for bottom slamming and at least 10 frame stations for bow flare and stern slamming are to be used in the slamming calculations for each of the bow and the stern regions.

4.2 Bottom Slamming

The bottom slamming pressure is to be calculated for the bottom structure in way of the flat of bottom forward of $0.25L$ measured from the forward perpendicular (FP), as shown in Fig. 4.1. Locations with local body plan angle less than $6°$ are to be investigated for bottom slamming. Typical frame stations and panel locations for bottom slamming are shown in Fig. 4.2.

4.3 Bow Flare Slamming

The bow flare slamming pressure is to be calculated for the side shell structure above the waterline in the area forward of $0.25L$ from the FP, as shown in Fig. 4.1. Typical frame stations and panel locations for bow flare slamming are shown in Fig. 4.2.

© The Author(s), under exclusive license to Springer Nature Switzerland AG 2025   13
F. Karkori, *Ship Vibration 4*, Synthesis Lectures on Ocean Systems Engineering,
https://doi.org/10.1007/978-3-031-74766-3_4

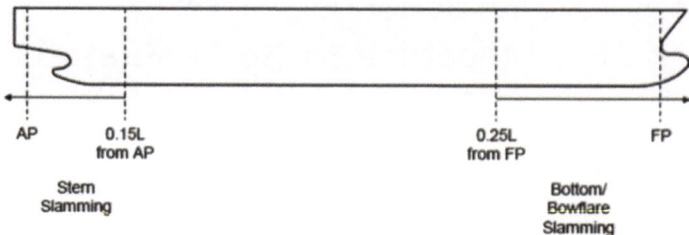

Fig. 4.1 Extent of hull structure for slamming load prediction

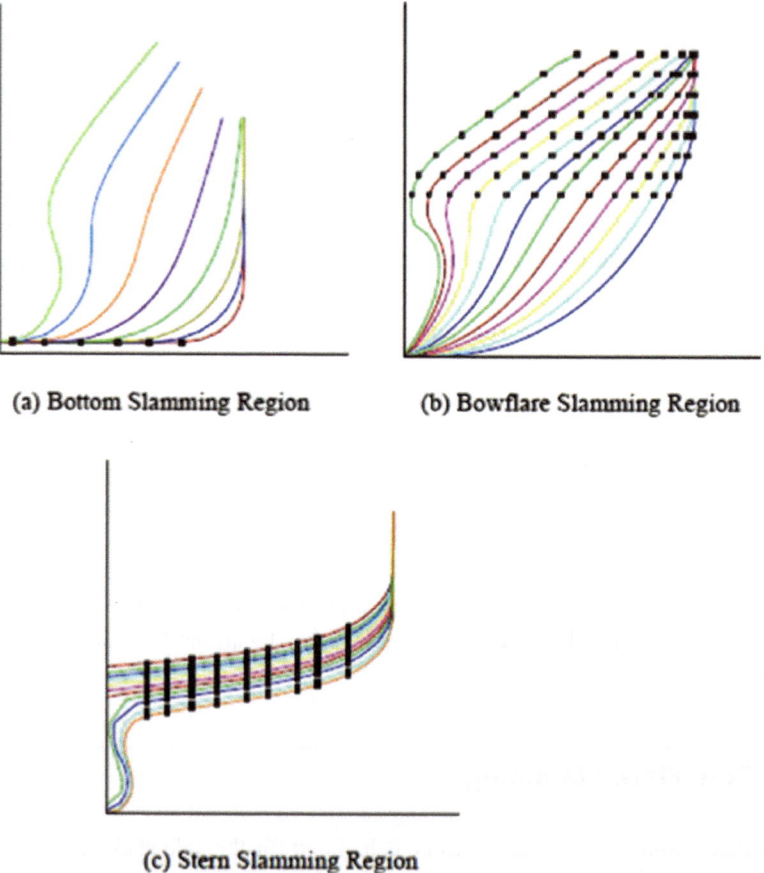

(a) Bottom Slamming Region   **(b) Bowflare Slamming Region**

(c) Stern Slamming Region

Fig. 4.2 Typical frame stations and panel locations for bow flare and stern slamming load prediction

4.4   Stern Slamming

The stern slamming pressure is to be calculated for the shell structure between $0.15L$ from the after perpendicular (AP) and the aft end, as shown in Fig. 4.1. Overhang areas and areas close to centrelines are critical due to the relatively shallow entry angle of the stern cross sections. Typical frame stations and panel locations for stern slamming are shown in Fig. 4.2.

Wave Environments

5

5.1 General

As seagoing vessels are typically designed for unrestricted service in the North Atlantic, the wave scatter diagram from *IACS Recommendation 34* may be employed for direct strength assessment against slamming loads. Table 5.1 shows the wave scatter diagram, where T_z is the average zero up-crossing wave period and H_{sis}, the significant wave height. The numbers in the diagram represent the probability of sea states described as occurrences per 100,000 observations.

5.2 Wave Spectrum

Sea wave conditions are to be modelled by the two-parameter Bretschneider spectrum, which is determined by the significant wave height and the zero-crossing wave period of a sea state. The wave spectrum is:

$$S_\zeta(\omega) = \frac{5\omega_p^4 H_s^2}{16\omega^5} \exp\left[-1.25(\omega_p/\omega)^4\right]$$

where

S_ζ wave energy density, m^2-s (ft^2-s)
H_s significant wave height, m (ft)
ω angular frequency of wave component, rad/s
ω_p peak frequency, rad/s
  $2\pi/T_p$

© The Author(s), under exclusive license to Springer Nature Switzerland AG 2025
F. Karkori, *Ship Vibration 4*, Synthesis Lectures on Ocean Systems Engineering,
https://doi.org/10.1007/978-3-031-74766-3_5

Table 5.1 IACS recommendation 34 wave scatter diagram for the North Atlantic

Hs (m)	T_z (s)																Sum
	3.5	4.5	5.5	6.5	7.5	8.5	9.5	10.5	11.5	12.5	13.5	14.5	15.5	16.5	17.5	18.5	
0.5	1.3	133.7	865.6	1186.0	634.2	186.3	36.9	5.6	0.7	0.1	0.0	0.0	0.0	0.0	0.0	0.0	3050
1.5	0.0	29.3	986.0	4976.0	7738.0	5569.7	2375.7	703.5	160.7	30.5	5.1	0.8	0.1	0.0	0.0	0.0	22575
2.5	0.0	2.2	197.5	2158.8	6230.0	7449.5	4860.4	2066.0	644.5	160.2	33.7	6.3	1.1	0.2	0.0	0.0	23810
3.5	0.0	0.2	34.9	695.5	3226.5	5675.0	5099.1	2838.0	1114.1	337.7	84.3	18.2	3.5	0.6	0.1	0.0	19128
4.5	0.0	0.0	6.0	196.1	1354.3	3288.5	3857.5	2685.5	1275.5	455.1	130.9	31.9	6.9	1.3	0.2	0.0	13289
5.5	0.0	0.0	1.0	51.0	498.4	1602.9	2372.7	2008.3	1126.0	468.6	150.9	41.0	9.7	2.1	0.4	0.1	8328
6.5	0.0	0.0	0.2	167.0	690.3	1257.9	1268.6	825.9	386.8	140.8	42.2	10.9	2.5	2.5	0.5	0.1	4806
7.5	0.0	0.0	0.0	3.0	52.1	270.1	594.4	703.2	524.9	276.7	111.7	36.7	10.2	2.2	0.6	0.1	2586
8.5	0.0	0.0	0.0	0.7	15.4	97.9	255.9	350.6	296.9	174.6	77.6	27.7	8.4	1.7	0.5	0.1	1309
9.5	0.0	0.0	0.0	0.2	4.3	33.2	101.9	159.9	152.2	99.2	48.3	18.7	6.1	1.2	0.4	0.1	626
10.5	0.0	0.0	0.0	0.0	1.2	10.7	37.9	67.5	71.7	51.5	27.3	11.4	4.0	0.7	0.3	0.1	285
11.5	0.0	0.0	0.0	0.0	0.3	3.3	13.3	26.6	31.4	24.7	14.2	6.4	2.4	0.4	0.2	0.1	124
12.5	0.0	0.0	0.0	0.0	0.1	1.0	4.4	9.9	12.8	11.0	6.8	3.3	1.3	0.2	0.1	0.0	51
13.5	0.0	0.0	0.0	0.0	0.0	0.3	1.4	3.5	5.0	4.6	3.1	1.6	0.7	0.1	0.1	0.0	21
14.5	0.0	0.0	0.0	0.0	0.0	0.1	0.4	1.2	1.8	1.8	1.3	0.7	0.3	0.1	0.0	0.0	8
15.5	0.0	0.0	0.0	0.0	0.0	0.0	0.1	0.4	0.6	0.7	0.5	0.3	0.1	0.1	0.0	0.0	3
16.5	0.0	0.0	0.0	0.0	0.0	0.0	0.0	0.1	0.2	0.2	0.2	0.1	0.1	0.0	0.0	0.0	1
Sum	1	165	2091	9280	19922	24879	20870	12898	6245	2479	837	247	66	16	3	1	100000

T_p peak period, sec
   $1.408 \, T_z$

To consider short-crested waves, the "cosine squared" spreading is to be utilised, which is defined as:

$$f(B) = k\cos^2(\beta - \beta_0)$$

where

β wave heading, following sea is 0, and head sea is 180°, in the range of
   $\beta_0 - \frac{\pi}{2} \leq \beta \leq \beta_0 + \frac{\pi}{2}$

β_0 main wave heading of a short-crested wave

k a factor determined such that summation of $f(\beta)$ is equal to 1.0, i.e.:
   $\sum_{\beta_0 - \pi/2}^{\beta_0 + \pi/2} f(\beta) = 1$

Vessel Motions

6

6.1 General

Vessel motions can be obtained through seakeeping analysis. The motions of interest for slamming load prediction are relative velocity and relative motion in the vertical direction between waves and the part of the vessel subject to wave impact.

A three-dimensional seakeeping code is to be used for the intended seakeeping analysis. The seakeeping analysis includes mainly the prediction of Response Amplitude Operators (RAOs) of the relative velocity and relative motion.

6.2 Response Amplitude Operator

The Response Amplitude Operators (RAOs) are the vessel's responses to unit amplitude, regular, sinusoidal waves. Computations of the ship motion RAOs are to be carried out using a linear seakeeping analysis code utilizing three-dimensional potential flow-based diffraction-radiation theory. All six degrees-of-freedom rigid-body motions of the vessel are to be accounted for. The RAOs are to be calculated for a number of wave headings and frequencies. Combined with wave spectrum, RAOs are to be used to determine extreme values for the duration of interest.

For the slamming load prediction, the RAOs of relative vertical velocities and motions are to be calculated at the pressure panel locations, as shown in Chap. 3, Fig. 3.2.

A sufficient range of wave headings and frequencies is to be considered for the calculation of the RAOs (refer to Chap. 2). It is recommended that the RAOs be calculated in heading increments of 15°.

The range of wave frequencies is to include 0.2–1.8 rad/s in increments of 0.05 rad/s. As specified in Chap. 2), a standard speed profile is employed for vessel speed.

© The Author(s), under exclusive license to Springer Nature Switzerland AG 2025   21
F. Karkori, *Ship Vibration 4*, Synthesis Lectures on Ocean Systems Engineering,
https://doi.org/10.1007/978-3-031-74766-3_6

Fig. 6.1 Panel arrangement
for seakeeping analysis

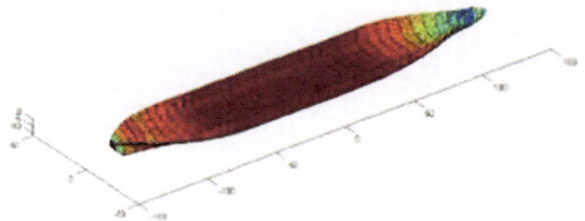

6.2.1 Seakeeping Model Development

Linear three-dimensional seakeeping analysis, in general, requires that the wetted surface of the vessel be discretised into a sufficiently large number of panels. Panels are to be generated on the vessel's surface as shown in Fig. 6.1. The panel mesh is to be fine enough to resolve the radiation and diffraction waves with reasonable accuracy.

6.2.2 Roll Damping

The roll motion of a vessel in beam or oblique seas is greatly affected by viscous roll damping, especially with wave frequencies near the roll resonance. For seakeeping analysis based on potential flow theory, a proper viscous roll damping model is required. Experimental data or empirical methods can be used for the determination of the viscous roll damping. In addition to the hull viscous damping, the roll damping due to bilge keels and anti-roll fins, where fitted, is to be considered. In absence of the required roll damping data, 10% of the critical damping may be used for overall viscous roll damping.

Motion Statistics 7

7.1 General

In this chapter, a short-term approach for statistical analysis of relative motion and velocity is presented. The objective is to obtain the lifetime maximum extreme values of the relative velocity and motion to be used for the slamming pressure calculation.

7.2 Short-Term Approach

7.2.1 Sea States

In order to consider a target service life, a series of sea states is selected from the IACS scatter diagram for the North Atlantic Ocean. The selection of sea states is based on the probability of occurrence. Figure 7.1 shows 1-year, 25-year, and 40-year return sea states selected from the wave scatter diagram. Table 7.1 shows the significant wave heights from the sea states with different return periods. Alternatively, a long-term approach, in which all the sea states in the scatter diagram are accounted for, may be employed.

7.2.2 Spectral Moment of Response

For each sea state the n-th spectral moment, m_n, of the response under consideration may be calculated, within the scope of linear theory, from the following equation:

$$m_n = \int_0^\infty \omega_e^n S_y(\omega_e) d\omega_e$$

© The Author(s), under exclusive license to Springer Nature Switzerland AG 2025
F. Karkori, *Ship Vibration 4*, Synthesis Lectures on Ocean Systems Engineering,
https://doi.org/10.1007/978-3-031-74766-3_7

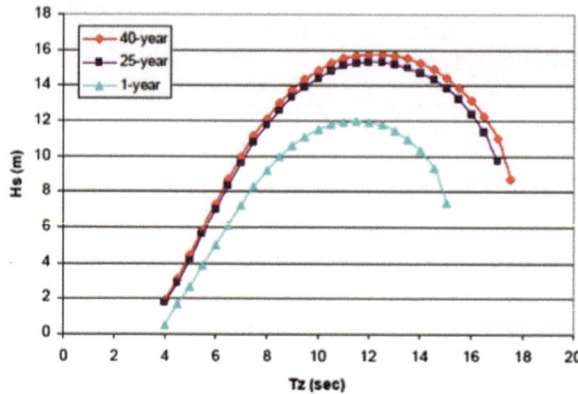

Fig. 7.1 1-, 25-, 40-Year return sea states from IACS recommendation 34 wave scatter diagram

Table 7.1 Sea states derived from IACS recommendation 34 wave scatter diagram

T_z (s)	H_s (m)				
	1-Year	20-Year	25-Year	30-Year	40-Year
4.0	0.5	1.7	1.7	1.8	1.9
4.5	1.6	2.8	2.9	3.0	3.1
5.0	2.7	4.1	4.2	4.3	4.4
5.5	3.8	5.5	5.6	5.7	5.9
6.0	5.0	6.9	7.0	7.1	7.3
6.5	6.2	8.2	8.4	8.5	8.7
7.0	7.3	9.5	9.6	9.8	10.0
7.5	8.3	10.6	10.8	10.9	11.1
8.0	9.2	11.6	11.8	11.9	12.1
8.5	10.0	12.5	12.6	12.8	13.0
9.0	10.6	13.2	13.4	13.5	13.8
9.5	11.1	13.8	14.0	14.1	14.4
10.0	11.5	14.3	14.5	14.6	14.9
10.5	11.8	14.6	14.8	15.0	15.2
11.0	12.0	14.9	15.1	15.3	15.5
11.5	12.0	15.1	15.3	15.4	15.7

where

m_n	n-th spectral moment at β_0 main heading angle.
$S_y(\omega_e)$	spectral density function of the response

$$\int_0^{2\pi} |H(\omega_e, \beta)|^2 S_\zeta(\omega_e, \beta)d\beta$$

ω_e	encounter frequency of vessel motion

$$\left|\omega - V\frac{\omega^2}{g}\cos\beta\right|$$

$S_\zeta(\omega_e\beta)$	directional wave spectrum at heading angle $.\beta$
$H(\omega_e\beta)$	Response Amplitude Operator (RAO) of the response at heading angle, β
β	wave heading angle.
g	gravitational acceleration.
V	vessel speed.

Using variance-preserving transformation, the equation for spectral moment can be written in wave frequency, ω, as follows:

$$m_n = \int_0^\infty \int_0^{2\pi} \omega_e^n |\omega, \beta|^2 S_\zeta(\omega, \beta)d\beta d\omega$$

Considering the short-crested waves specified in Chap. 4, this equation can be written as follows:

$$m_n = \sum_{\beta_0-\frac{\pi}{2}}^{\beta_0-\frac{\pi}{2}} f(\beta)\left[\int_0^\infty \omega_e^n |H(\omega, \beta)|^2 S_\zeta(d)d\omega\right]$$

where

f	spreading function.
β	wave heading, following sea is 0, and head sea is $180°$, in the range of $\beta_0 - \frac{\pi}{2} \le \beta \le \beta_0 + \frac{\pi}{2}$
β_0	main wave heading angle of a short-crested wave

7.2.3 Short Term Extreme Values

Assuming that the wave-induced response is a Gaussian stochastic process with zero mean and the spectral density function $S_y(\omega)$ is narrow banded, the probability density function of the maxima (peak values) may be represented by a Rayleigh distribution. Then, the short-term probability of the response exceeding x_0, $Pr\{x_0\}$ for the j-th sea state may be expressed by the following equation:

$$Pr_j(x_0) = \exp\left(-\frac{x_0^2}{2m_0^j}\right)$$

Based on Rayleigh distribution, the extreme values of relative vertical velocity, v, can be calculated by the following equation:

$$v = \sqrt{2\sigma_v^2\left(\ln\frac{t}{T_2}\right)} \qquad \text{for bow flare slamming}$$

$$= \sqrt{2\sigma_v^2\left[\ln\left(\frac{t}{T_2}\right) - \frac{d_i^2}{2\sigma_r^2}\right]} \text{ for bottom and stern slamming}$$

where

v relative vertical velocity, in m/s (ft/s)
σ_v standard deviation of relative vertical velocity, in m/s (ft/s)
σ_r standard deviation of relative vertical motion, in m (ft)
d_i vertical distance from the still water surface to the location, in m (ft), if the location is above the water surface, then it becomes zero
t total duration time, in seconds, 3 h in storm time
T_2 mean period of relative vertical velocity, in seconds
$2\pi\sqrt{\frac{m_0}{m_2}}$

The standard deviation is the square root of the 0-th order of the response spectral moment. A short-term spectral analysis may be performed using COTS or Class developed computer programmes.[1]

7.2.4   Lifetime Maximum Extreme Values

According to the short-term approach, the lifetime maximum extreme values are estimated by finding the maximum of the short-term extreme values for a three-hour duration from the 25-year return sea states. This estimation is based on the assumption that the 25-year return sea states have approximately 10-5 probability of occurrence, and the most probable short-term extreme values for three-hour duration have about 10-3 probability of exceedance. Combining these two probabilities, the extreme values in the 25-year return sea states are representative of a 10-8 probability level.

---

[1] For example, *SPECTRO* from ABS.

7.3 Long-Term Approach

The lifetime maximum extreme values may be also calculated from the long-term approach by the joint probability of short-term extreme values and occurrence of sea states in the wave scatter diagram.

Slamming Pressure

8

8.1 General

The slamming pressure can be obtained from either experiments or numerical analyses. This chapter is focused on the two-dimensional (2D) slamming analysis using short-term extreme velocities and local pressure coefficients, which is called a 2D slamming approach. An advanced approach may also be applied to obtain the slamming pressure, such as three-dimensional (3D) slamming model tests or Computational Fluid Dynamics (CFD) simulations, which is called a 3D slamming approach. In this book, it is mainly focused on the 2D slamming approach.

8.2 2D Slamming Approach

This chapter introduces the 2D slamming approach to calculate the slamming pressures as a simplified design method. When the 2D slamming approach is employed, the slamming pressure is to be adjusted for the three-dimensional effects.

8.2.1 Slamming Pressure Coefficient

The slamming pressure coefficient is heavily influenced by the local geometry of a vessel such as bow or stern flare angle and deadrise angle. Three-dimensional effects are also important in the determination of the slamming pressure. The slamming pressure coefficients at locations under consideration can be calculated using a computer programme based on a potential theory. The program is to predict the local slamming pressure and integrated force on a vessel section penetrating the calm water surface. The initial location

© The Author(s), under exclusive license to Springer Nature Switzerland AG 2025   29
F. Karkori, *Ship Vibration 4*, Synthesis Lectures on Ocean Systems Engineering,
https://doi.org/10.1007/978-3-031-74766-3_8

of the calm water surface in the 2D potential calculation is recommended to be the draft line for bow flare slamming and the bottom of the section or lower for bottom or stern slamming.

An acceptable computer program is to be formulated in a time domain solving a 2D hydrodynamic impact boundary value problem at each time step. The disturbance of the free surface due to the presence of the body is to be considered based on a suitable theory such as Wagner's theory of wave-rise. The programme is to require input data of the hull section geometry, initial displacement, and time history of the impact velocity. The input offset points are to be sufficient to capture the curvature of the hull section. The program is to calculate the total force, and average pressure over each panel. Depending on the structural configuration, the shell plating under consideration can be subdivided into panels for slamming pressure calculations. The panel length may be about four (4) frame spacing, which is about 3 m, suitable for calculating the equivalent static slamming pressures. The maximum of the average pressure over each panel during the slamming simulation, as shown in Figs. 8.1 and 8.2, is to be selected to determine the pressure coefficient at locations under consideration.

The slamming pressure coefficient for each location can be calculated using the following equation:

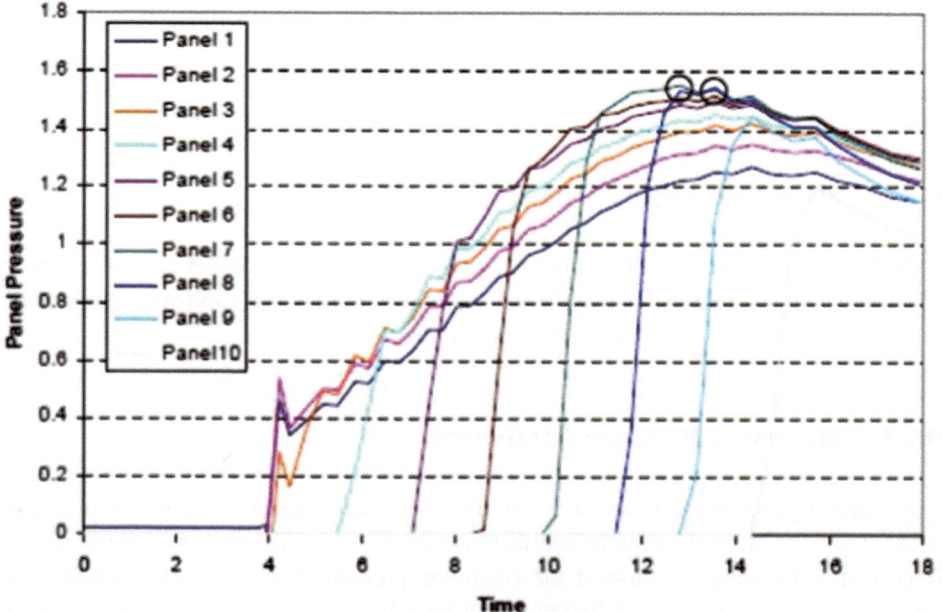

Fig. 8.1 Typical pressure time history in bow flare slamming

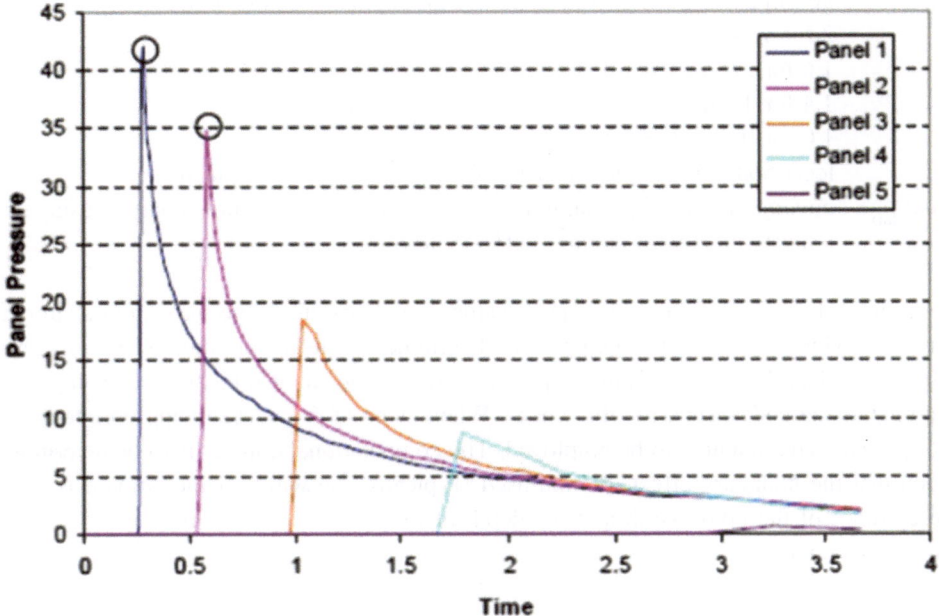

Fig. 8.2 Typical pressure time history in stern and bottom slamming

$$C_p = \frac{(\bar{p})_{\max}}{0.5kv^n}$$

where

v	vertical impact velocity used in 2D slamming analysis, in m/s (ft/s)
n	2
$(\bar{p})_{\max}$	maximum of the average panel pressure from 2D slamming analysis, in N/cm^2 (kgf/cm^2, lbf/in^2)
k	0.1025 (0.01045, 0.0138)

If a unit impact velocity is used for experimental or numerical slamming analysis, due consideration is to be given to the determination of the slamming pressure coefficient in conjunction with a pressure–velocity relationship. For the case of 2D slamming analysis, the following pressure–velocity relationship may be used for the calculation of slamming pressure coefficient:

$$C_p = \frac{(\bar{p}_0)_{\max}}{0.5kv_0^n}$$

where

v_0 unit impact velocity used in 2D slamming analysis, in m/s (ft/s)
n 2 for $\alpha_b \geq 6$
           1.6 for $3 \leq \alpha_b < 6$
           1.4 for $1 \leq \alpha_b < 3$
           1.0 for $0 \leq \alpha_b < 1$
α_b local body plan angle measured from the horizontal, in degrees
$(\bar{p}_0)_{max}$ maximum of average slamming pressure from 2D slamming analysis using unit
           impact velocity, in N/cm^2 (kgf/cm^2, lbf/in^2)

For each selected panel, there is a peak value corresponding to the maximum slamming pressure. This maximum average pressure determines the local maximum slamming pressure coefficient. The local slamming pressure coefficient can also be obtained using other computational codes, such as CFD codes. When using CFD codes, an appropriate model for pressure coefficient is to be employed. The computational domain is to be determined such that the boundary effect is minimised. Typically, the width of the domain is about three times of the body width and the depth of the domain is larger than three times of the body depth.

8.2.2 Three-Dimensional Effects

To consider three dimensional effects, a correction factor is to be used. For bottom slamming near FP, a location factor is to be multiplied to C3D.

$$C_{3D} = \text{three-dimensional correction factor}$$
$$= 0.83 \quad \text{for bow flare and stern slamming}$$
$$= 0.83 \quad \text{for bottom slamming}$$

where

C_L location factor for bottom slamming
       1.0 at 0.25 L from FP
       1.0 at $[0.1 - 0.5(C_b - 0.7)L$ from FP
       0.5 at and forward of FP

Linear interpolation is to be used for the intermediate longitudinal locations.

8.2.3 Dynamic Effects

The slamming pressure is a time dependent function characterized by rise time and duration. Figures 8.1 and 8.2 depict typical time histories of dynamic pressure from the two-dimensional slamming analysis for bow flare slamming and bottom/stern slamming, respectively. The pressure points are similar to those shown in Fig. 8.2. The design slamming pressure for direct strength assessment is an equivalent static pressure that would cause the same maximum structural response as the dynamic slamming pressure. The design slamming pressure is a function of the dynamic load factor. The dynamic load factor C_s can be obtained by performing structural analysis applying dynamic load and static load. If such analysis is not available, the following dynamic load factor can be used.

$$C_s = f_1 \omega_1 \sqrt{L} = \text{for } \alpha_b < 6$$

but need not to be taken greater than for in m

$$(0.0201(360 - 0.1L)^{\frac{1}{2}} \text{ for in ft})$$

$$C_s = 1 \qquad \text{for } 30 \leq \alpha_b$$

Linear interpolation is to be used for intermediate local body plan angles.
Where

ω_1 2-node hull girder vertical vibratory frequency in rad/sec. If not known, the following formula by Kumai may be used:

$$\mu \left[I_v / \Delta_i L^3 \right]^{\frac{1}{2}} \text{ in rad/s}$$

I_v moment of inertia amidships, in m^4 (ft^4).

Δ_i virtual displacement, including added mass of water, in tons (L tons)

$$\left[1.2 + \frac{B}{3d_m} \right] \Delta$$

Δ vessel displacement, in tons (L tons)

μ 321,500 (176,100)

B breadth of vessel, in m (ft)

d_m mean draft of vessel, in m (ft)

f_1 0.004 (0.0022)

L scantling length, in m (ft)

α_b local body plan angle measured from the horizontal, in degrees

8.2.4 Design Slamming Pressure

For each location, the design slamming pressure, p_s, can be calculated using the following equations:

$$p_s = \frac{1}{2}kC_sC_{3D}C_pv^n \quad \text{in N/cm}^2\left(\text{kgf/cm}^2, \text{ lbf/in}^2\right)$$

where

k 0.1025 (0.01045, 0.148).

C_s dynamic load factor.

C_{3D} three-dimensional correction factor.

C_P local pressure coefficient.

n a constant for pressure–velocity relationship.

v relative vertical velocity between water and hull, in m/s (ft/s), calculated from the statistical analysis.

For each location, the largest slamming pressure from the 25-year return sea states is selected as the design slamming pressure.

8.3   3D Slamming Approach

The 3D slamming approach is another option based on the three-dimensional slamming analysis considering the fully nonlinear ship motions in sea states or equivalent design waves. The 3D slamming approach may take significant computational efforts and may be used as a higher-level approach for evaluation of critical conditions.

Direct Strength Assessment

9

9.1 General

The criteria in this chapter, except those for main supporting members, are based on a "net" ship approach wherein the nominal design corrosion values are deducted. The nominal design corrosion values for plating and structural members are to be applied, as shown in Table 9.1.

9.2 Simultaneous Load Factors

In the direct finite element-based strength assessment of main supporting members, the simultaneous application of the design maximum pressure is deemed to be too conservative. This is because the design pressure is the envelope of the local maximum pressure. To account for simultaneous loading of these maximum pressures on the finite element model, the spatial distribution of instantaneous slamming pressures on the region of interest of the hull is to be expressed by multiplying the calculated maximum slamming pressure by a factor, shown in Table 9.2 depending on the region of interest.

9.3 Shell Plating

9.3.1 Bow Flare Slamming

The net thickness of the shell plating is not to be less than t_1 or t_2, whichever is greater, obtained from the following equations:

© The Author(s), under exclusive license to Springer Nature Switzerland AG 2025   35
F. Karkori, *Ship Vibration 4*, Synthesis Lectures on Ocean Systems Engineering,
https://doi.org/10.1007/978-3-031-74766-3_9

Table 9.1 Nominal design corrosion values for direct strength assessment

	Nominal design corrosion values, in mm (in)
Shell plating	1.0 (0.04) in way of void space 1.0 (0.04) in way of ballast tank space (below the upper turn of the bilge) 1.5 (0.06) in way of ballast tank space (above the upper turn of the bilge)
Web and flange of stiffeners and main supporting members	1.0 (0.04) in way of void space 1.5 (0.06) in way of ballast tank space (vertical members) 2.0 (0.08) in way of ballast tank space (horizontal members)

Table 9.2 Simultaneous load factors for finite element analysis of main supporting

Type	Factor for simultaneous loading
Bottom slamming	0.4
Bow flare slamming	0.71
Stern slamming	0.5

$$t_1 = 0.73s(k_1 p_s/f_2)^{\frac{1}{2}} \text{ mm (in)}$$

$$t_2 = 0.73s(k_2 p_s/f_2)^{\frac{1}{2}} \text{ mm (in)}$$

where

s spacing of longitudinal or transverse frames, in mm (in)

k_1 0.342   for longitudinally stiffened plating

       0.5k2   for transversely stiffened plating

k_2 0.5   for longitudinally stiffened plating

       0.342   for transversely stiffened plating

k $\left[\dfrac{3.075(\alpha)^{\frac{1}{2}} - 2.077}{\alpha + 0.272} \right] (1 \le \alpha \le 2)$

     $= 1.0 \quad \alpha > 2$

α aspect ratio of the panel (longer edge/shorter edge)

p_s slamming pressure, in N/cm^2 (kgf/cm^2, lbf/in^2)

f_1 0.90 $S_m f_y$ for side shell plating in the region forward of 0.125 L from the forward perpendicular, in N/cm^2 (kgf/cm^2, lbf/in^2)

     0.75 $S_m f_y$ for side shell plating in the region between 0.125 and 0.25 L, from the forward perpendicular

f_2 $0.95\ S_m f_y$ in N/cm² (kgf/cm², lbf/in²)

S_m strength reduction factor

    1   for Ordinary Strength Steel

    0.95 for Grade H32 Steel

    0.908   for Grade H36 Steel

    0.875   for Grade H40 Steel

f_y minimum specified yield point of the material, in N/cm² (kgf/cm², lbf/in²)

9.3.2   Bottom and Stern Slamming

In the case of bottom or stern slamming, an ultimate strength approach considering plasticity may be employed to estimate plate thickness to the corresponding slamming pressure. The net thickness of the hull envelope plating, t, is not to be less than:

$$t = \frac{k_1 \alpha_p s}{c_d} \sqrt{\frac{p_s}{c_a f_y}} \quad \text{mm (in)}$$

where

k_1 0.5

α_p correction factor for the panel aspect ratio.

    $1.2 - \frac{s}{k_2 \ell}$ but not to be taken as greater than 1.0

k_2 2100 (2100, 25.2)

s spacing of longitudinal or transverse frames, in mm (in)

ℓ unsupported span of the frame, in m (ft)

p_s design slamming pressure, in N/cm² (kgf/cm², lbf/in²)

C_d plate capacity correction coefficient

    1.0

C_a permissible bending stress coefficient

    1.0

f_y minimum specified yield point of the material, in N/cm² (kgf/cm², lbf/in²)

9.4   Shell Longitudinals and Stiffeners

9.4.1   Bow Flare Slamming

The net section modulus of the shell longitudinal (or frame), including the associated effective plating, is not to be less than that obtained from the following equation:

$$SM = M/f_b \quad \text{cm}^3/(\text{in}^3)$$

$$M = \frac{p_s s \ell^2 10^3}{k} \quad \text{N-cm (kgf-cm, lbf-in)}$$

where

k 16 (16, 111.1)

p_s design slamming pressure, in N/cm² (kgf/cm², lbf/in²)

s spacing of longitudinal or transverse frames, in mm (in.)

ℓ unsupported span of the frame, in m (ft)

p_s $0.9 S_m f_y$ for transverse and longitudinal frames in the region forward of 0.125 L from the FP, in N/cm² (kgf/cm², lbf/in²), but $0.8 S_m f_y$ for longitudinal frames above 0.85 D.

       $0.8 S_m f_y$ for transverse and longitudinal stiffeners in the region between 0.125 L and 0.25 L, measured from the FP, in N/cm² (kgf/cm², lbf/in²), but $0.7 S_m f_y$ for longitudinal stiffeners above 0.85D

       $0.9 S_m f_y$ for stern slamming, in N/cm² (kgf/cm², lbf/in²)

The associated effective breadth of plating is to be taken as spacing of longitudinal or transverse frames or 20% of the unsupported span, whichever is less. In addition to the above, the net web thickness of shell longitudinals is not to be less than obtained from the following equation:

$$t_{w\_req} = 0.5 \frac{p_s s \ell_{shr}}{d_{shr} \tau_a}$$

where

τ_a $0.4 S_m f_y$

ℓ_{shr} effective shear span, in m (ft)

$d\ell_{shr}$ effective shear depth of stiffener, in mm (in)

9.4.2   Bottom and Stern Slamming

In case of bottom or stern slamming, an ultimate strength approach considering plasticity may be employed to estimate section modulus to the corresponding slamming pressure. The net plastic section modulus, SM_{pl}, of each individual stiffener, is not to be less than:

$$SM_{pl} = \frac{k p_s s \ell^2}{f_{bdg} C_a f_y} \quad \text{cm}^3 \text{(in}^3\text{)}$$

where

k	1000 (1000, 144).
p_s	slamming pressure, in N/cm² (kgf/cm², lbf/in²).
s	spacing of longitudinal or transverse frames, in mm (in).
ℓ	unsupported span of the frame, in m (ft).
f_{bdg}	bending moment factor

$$8\left(1 + \tfrac{n_s}{2}\right)$$

n_s	2.0 for continuous stiffeners or where stiffeners are bracketed at both ends
C_a	permissible bending stress coefficient.
	0.9
F_y	minimum specified yield point of the material, in N/cm² (kgf/cm², lbf/in²).

The associated effective breadth of plating may be taken as the spacing of longitudinal or transverse frames. The net plastic section modulus can be calculated using the following formulae. When the cross-sectional area of the attached plate exceeds the cross-sectional area of the stiffener to which the plate flange is attached, the actual net plastic section modulus, z_p, in cm³ (cm³, in³) is given by:

$$Z_p = A_{pn}t_{pn}(2c_4) + \frac{h_w^2 t_{wn} sin\varphi_w}{2 \cdot c_4^3} + A_{fn}(h_{fc} sin\varphi_w - b_w)/c_4$$

where

A_{pn}	net cross-sectional area of the attached plate, in cm² (cm², in²)
t_{pn}	net attached plate thickness, in mm (mm, in)
h_w	height of stiffener web, in mm (mm, in), refer to Fig. 9.1
b_w	distance from mid thickness plane of stiffener web to the center of the flange area, in mm (mm, in.), refer to Fig. 9.1
c_4	10 (10, 1).
h	height of stiffener, in mm (mm, in.), refer to Fig. 9.1
t_{wn}	net web thickness, in mm (mm, in.)
	$t_w - t_c$
t_w	gross web thickness, in mm (mm, in.), refer to Fig. 9.1
t_c	corrosion deduction, in mm (mm, in.), to be subtracted from the web and flange thickness.
φ_w	smallest angle between attached plate and stiffener web, measured at the midspan of the stiffener, refer to Fig. 9.1.
	The angle φ_w may be taken as 90 degrees provided the smallest angle is not less than 75 degrees.
s	spacing of longitudinal or transverse frames, in m (m, in).

When the cross-sectional area of the stiffener exceeds the cross-sectional area of the attached plate, the plastic neutral axis is located a distance z_{na}, in mm (mm, in.), above

Fig. 9.1 Stiffener geometry

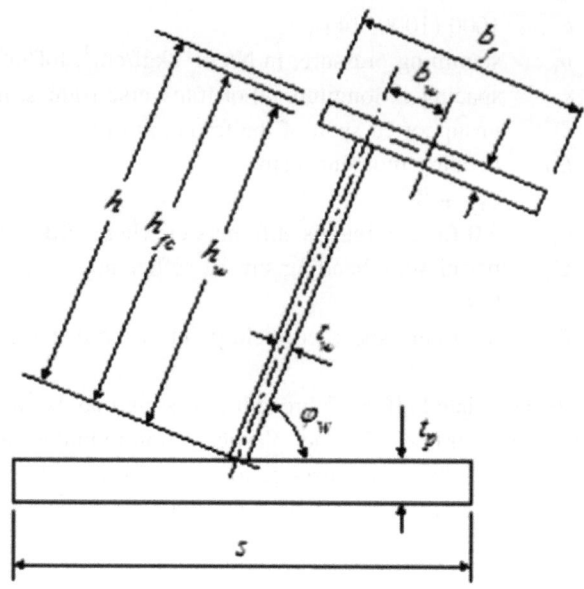

the attached plate, given by:

$$z_{na} = c_4^2 A_{fn} + h_w t_{wn} - c_4^3 t_{pn}s)/(2t_{wn})$$

and the net plastic section modulus, z_p, in cm³ (cm³, in³) is given by:

$$Z_p = t_{pn}s(z_{na} + \frac{t_{pn}}{2)\sin\varphi_w}$$
$$+ \left[\frac{[h_w - z_{na}]^2 + z_{na}^2]t_{wn}\sin\varphi_w}{2 \cdot c_4^3} + A_{fn}[(h_{fc} - z_{na})\sin\varphi_w - b_w\cos\varphi_w]/c_4\right]$$

9.5   Slot Connections

Each slot connection under the design slamming pressure is to be verified using the following formulae:

$$\sigma_{fb} = P_1/A_s < S_m f_y$$

$$\tau_{dc} = \frac{P_2}{A_c} < \frac{S_m f_y}{\sqrt{3}}$$

where

σ_{fb} flat bar mean stress, in N/cm^2 (kgf/cm^2, lbf/in^2)

τ_{dc} direct collar plate mean stress, in N/cm^2 (kgf/cm^2, lbf/in^2)

P_1 load transmitted through flat bar stiffener, in N (kgf, lbf)

$$p_s s \ell \left(1 - \tfrac{1}{2}\alpha_s \left(\tfrac{4 f_c A_s}{4 f_c A_x + A_c} \right) - \alpha_s \right)$$
if the flat bar stiffener is connected to the longitudinal stiffener

0
if the flat bar stiffener is not connected to the longitudinal stiffener

P_2 load transmitted through shear connection, in N (kgf, lbf)

$$p_s s \ell \left(1 - \tfrac{1}{2}\alpha_s \left(\tfrac{A_c}{4 f_c A_s + A_c} \right) - \alpha_s \right)$$
if the flat bar stiffener is connected to the longitudinal stiffener

$$p_s s \ell \left(1 - \tfrac{1}{2}\alpha_s \right)$$
if the flat bar stiffener is not connected to the longitudinal stiffener

P_2 slamming pressure, in N/cm^2 (kgf/cm^2, lbf/in^2)

s spacing of longitudinal/stiffener, in cm (in)

ℓ spacing of transverses, in cm (in)

A_s net attached area of the flat bar stiffener, in cm^2 (in^2)

A_c effective net shear sectional area of the support or of both supports for double-sided support, in cm^2 (in^2)

$A_{lc} + A_{ld}$

A_{ld} net shear connection area excluding lug plate, in cm^2 (in^2)

$\ell_d t_{tw}$

ℓ_d length of direct connection between longitudinal stiffener and transverse member, refer to Fig. 9.2, in cm (in)

t_{tw} net thickness of transverse member (refer to Fig. 9.2), in cm (in)

A_{lc} net shear connection area of lug plate, in cm^2 (in^2)

$$f_1 \ell_c t_c$$

ℓ_c length of connection between longitudinal stiffener and lug plate (refer to Fig. 9.2), in cm (in)

t_c net thickness of lug plate (refer to Fig. 9.2), not to be taken greater than the thickness of adjacent transverse member, in cm (in)

f_1 shear stiffness coefficient

1.0   for stiffener of symmetrical cross - section

4/W 5.5/W = 1.0 for stiffener of asymmetrical cross - section

W width of the cut-out for an asymmetrical stiffener, measured from the cut-out side of the stiffener web (refer to Fig. 9.2), in cm (in)

f_c collar load factor

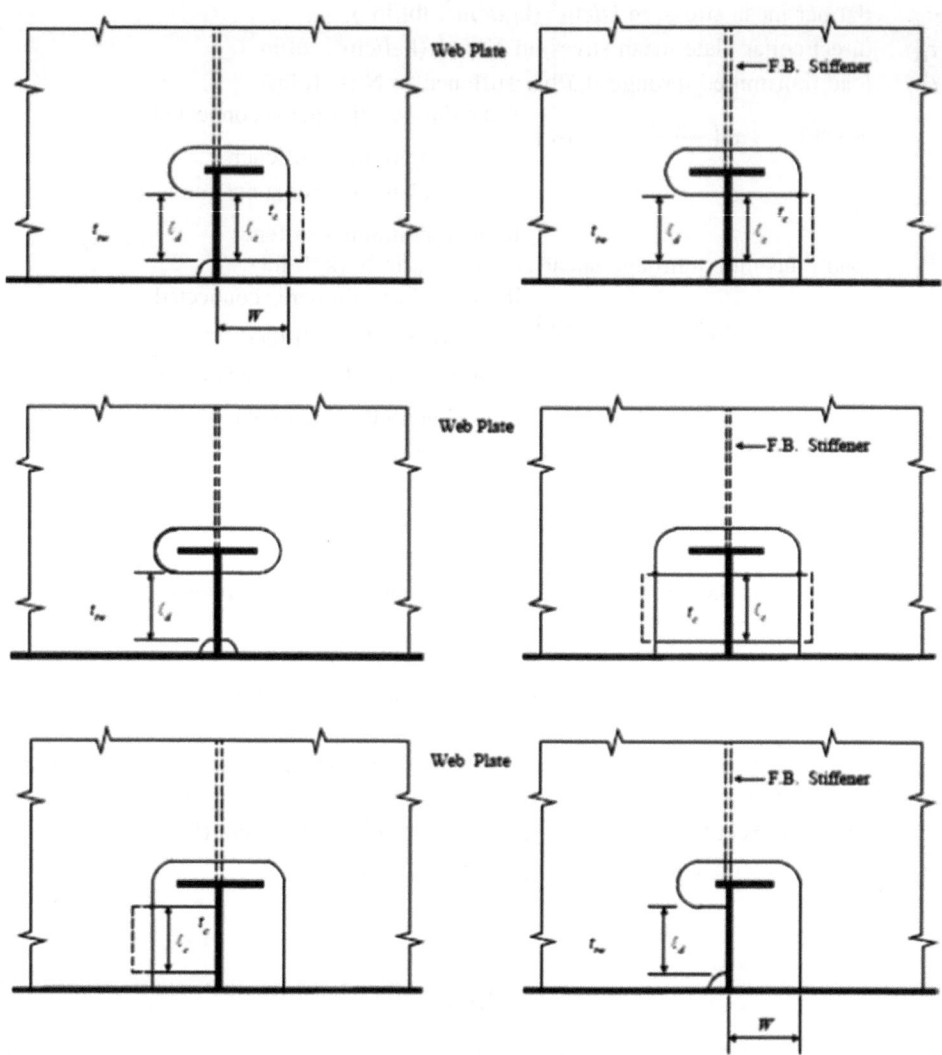

Fig. 9.2 Cut-outs (slots) for longitudinal

- *For intersecting of symmetrical stiffeners*
  for A_s in cm^2

1.85	for $A_s \leq 14$
$1.85 - 0.0441(A_s - 14)$	for $14 < A_s \leq 31$
$1.1 - 0.013(A_s - 31)$	for $31 < A_s \leq 58$
0.75	for $A_s > 58$

- for A_s in in^2

1.85   for $A_s \leq 2.2$

$1.85 - 0.2883(A_s - 2.2)$ for $2.2 < A_s \leq 4.8$

$1.1 - 0.0836(A_s - 4.8)$ for $4.8 < A_s \leq 9.0$

0.75   for $A_s > 9.0$

- *For intersecting of asymmetrical stiffeners*

  $0.68 + 0.0172\ell_d/A_s$ for ℓ_d in cm and A_s in cm^2

  $0.68 + 0.00677\ell_d/A_s$ for ℓ_d in inches and A_s in in^2

If the length of direct and shear connections are different, their mean value is to be used instead of ℓ_d, and in case of a single lug, the value is ℓ_c.

α_s panel aspect ratio, not to be taken greater than 0.25

$$\frac{s}{\ell}$$

s_m and f_y and are as defined in the section on *Shell plating*. For flat bar stiffener with soft-toed brackets, the brackets may be included in the calculation of A_s.

9.6   Main Supporting Members

For vessels having flat bottom and/or built with main supporting members, a finite element analysis is to be performed to verify the strength of the main supporting members under the slamming pressure from a direct analysis. In addition, transversely stiffened brackets connected to side shell longitudinal stiffeners and supported by horizontal stringers forming a structural grillage system subject to bow flare slamming loads are to be examined by grillage analysis or finite element analysis.

In addition to the finite element analysis for main supporting members, transverse web frames supporting these grillage structures are to be verified by nonlinear analysis with a simultaneous loading factor of 1.0 to reliably capture buckling and plastic deformation of panels in way of connections with side shell longitudinal stiffeners. For non-linear analysis, reference may be made to the Class Rules pertaining to nonlinear finite element analysis of marine and offshore structures, and the permanent lateral and out-of-plane deformation of the considered member are to be small relative to the relevant structural dimensions and within the "Standard" range in IACS recommendation No.47.

9.6.1   Structural Modelling

For bow flare and bottom slamming, the forebody structure forward of the aft transverse bulkhead of the foremost cargo hold is to be modelled. Generally, quadrilateral plate bending elements are used for the web plating of the main supporting members, and rod elements are used for flanges. However, in critical areas, flanges and stiffeners need to be modelled with plate elements. A global finite element mesh of one longitudinal spacing is to be applied with critical areas being modelled with finer mesh density. Critical members in the forebody region can be identified as: transverse frames with large openings, or large inclined angle and slot connections of side longitudinal stiffeners and transverse frames, etc. Gross scantlings are to be used for finite element modelling. Similar finite element modelling approach is applicable to the stern structure.

9.6.2   Boundary Constraints

To evaluate the slamming effects, the finite element model needs to be properly constrained. Figure 9.3 shows the boundary conditions that may be applied to the vessel forebody model. In the case of a symmetric structure, a half model can be used with symmetric boundary conditions applied on the centreline plane. To absorb and distribute any unbalanced forces in the vertical direction, vertical springs are to be placed along the outer shell and longitudinal bulkhead at transverse bulkheads (Line V in Fig. 9.2). These springs may be modelled using rod elements with the cross-sectional area defined as follows:

$$A_s = 0.77 \frac{A_{shear} \ell_s}{n \ell_t}$$

where
A_s cross-sectional area of the spring rod, in cm^2 (in^2)
A_{shear} effective shear area of a hull girder cross section, in cm^2 (in^2)

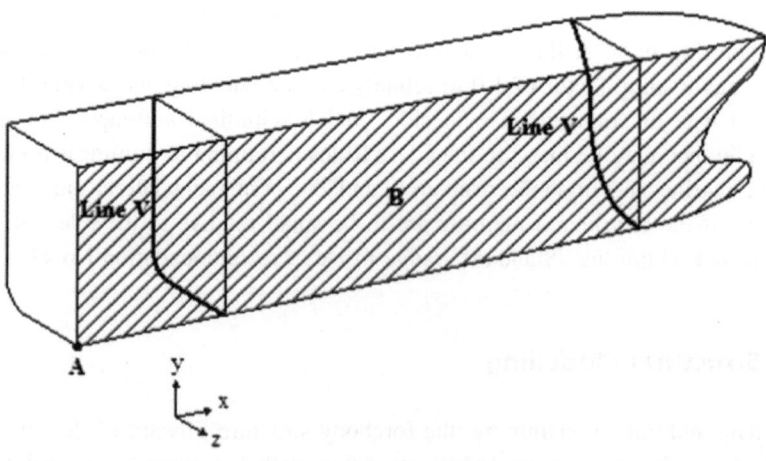

Fig. 9.3   Boundary constraints for forebody finite element model

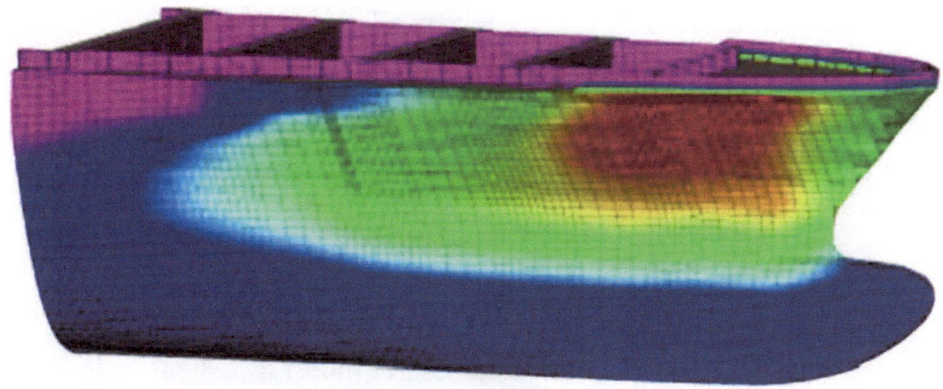

Fig. 9.4 Forebody slamming pressure mapped on finite element model

ℓ_s length of spring rod, in cm (in)

n number of spring rod elements

ℓ_t cargo hold length between transverse bulkheads, in cm (in)

Point A	UX = 0, UY = 0
Line V	Attach spring rod elements
Section B	UZ = 0, RX = 0, RY = 0

9.6.3 Slamming Pressure on FE Model

The design slamming pressure can be obtained from a direct calculation discussed in this chapter. The calculated pressure needs to be mapped on to the finite element model. Since the finite element analysis is being performed, the slamming pressure is adjusted by applying the simultaneous load factors given in Table 9.2. Figures 9.4 and 9.5 depict the mapped slamming pressures on a finite element model of a forebody and aft body, respectively.

9.6.4 Acceptance Criteria—Yielding

To assess the strength adequacy of main supporting members, the stresses in individual elements need to be checked against the yielding criteria. The allowable stresses defined in Table 9.3 are applicable to the main supporting members and structural details, except for slot connections. The allowable stress for the recommended basic mesh size is defined as

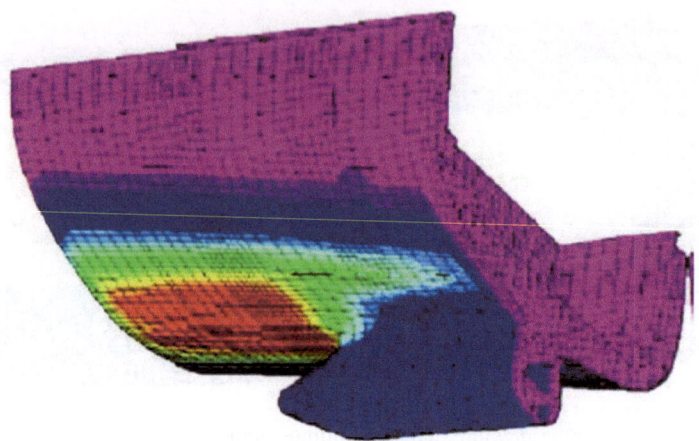

Fig. 9.5 Aft body slamming pressure mapped on finite element model

a percentage of the minimum specified yield stress f_y times the strength reduction factor s_m. Application of this stress to rod elements is based on axial stress while von-Mises membrane stresses for quadrilateral elements are checked. In Table 9.3, L_s is longitudinal spacing, and f_u is the minimum strength of the material. For the local stresses in way of slot connections, the allowable stresses are taken as 0.71 times the corresponding stress given in Table 9.3.

9.6.5 Acceptance Criteria—Buckling

In general, the stiffness of the web stiffeners along the depth of the web plating is to be in compliance with the following requirements.

The net moment of inertia, i, of the web stiffener, with the effective breadth of net plating not exceeding s or 0.33 ℓ, whichever is less, is not to be less than that obtained from the following equations:

$$i = 0.17\ell t^3 \left(\frac{\ell}{s}\right)^2 \mathrm{cm}^4 (\mathrm{in}^4) \quad \mathrm{for} \ell/s \leq 2.0$$

$$i = 0.34\ell t^3 \left(\frac{\ell}{s}\right)^2 \mathrm{cm}^4 (\mathrm{in}^4) \quad \mathrm{for} \ell/s \leq 2.0$$

where

ℓ length of stiffener between effective supports, in cm (in)
t required net thickness of web plating, in cm (in)

Table 9.3 Allowable stresses (kgf/cm^2) for various finite element mesh sizes

Mesh size	Stress limit	Mild steel ($s_m = 1.000$)	HT32 ($s_m = 0.950$)	HT36 ($s_m = 0.908$))
$1 \times LS$	$1.00 \times c_f^c S_m f_y$	2280	2888	3106
$1/2 \times LS^a$	$1.06 \times c_f S_m f_y$	2417	3061	3292
$1/3 \times LS^a$	$1.12 \times c_f S_m f_y$	2554	3234	3478
$1/4 \times LS^a$	$1.18 \times c_f S_m f_y$	2690	3408	3664
$1/5 \times LS \sim 1/10 \times LS^a$	$1.25 \times c_f S_m f_y$	2850	3610	3882
Thickness^{a,b}	$c_f f_u$ or^d $1.50 \times c_f S_m f_y$	3895	4275	4658

Notes
^{a} Stress limits greater than $1.00 \times c_f S_m f_y$ are to be restricted to small areas in way of structural discontinuities
^{b} When the fatigue strength of the detail is found satisfactory, the hot spot stress in the detail may be allowed up to the minimum tensile strength of the material
^{c} c_f is to be taken as 0.95
^{d} Smaller value is to be taken except for *Mild Steel* for which $c_f c_u$ is taken based on Note 2

s spacing of stiffeners, in cm (in)

Web stiffeners which are oriented parallel to and near the face plate, and thus subject to axial compression, are also to have adequate buckling strength, considering the combined effect of the compressive and bending stresses in the web. In this case, the unsupported span of these parallel stiffeners may be taken between tripping brackets, as applicable. The buckling strength of the web plate between stiffeners and flange/face plate is to satisfy the limit specified below:

$$\left(\frac{f_{LB}}{f_{cLb}}\right)^2 + \left(\frac{f_b}{f_{cb}}\right)^2 + \left(\frac{f_{LT}}{f_{cLT}}\right)^2 \le S_m$$

where

f_{LB}	calculated uniform compressive stress, in N/cm^2 (kgf/cm^2, lbf/in^2)
f_b	calculated ideal bending stresses, in N/cm^2 (kgf/cm^2, lbf/in^2)
f_{LT}	calculated total in-plane shear stress, in N/cm^2 (kgf/cm^2, lbf/in^2)
f_{cLb}, f_b and f_{LT}	are to be calculated for the panel in question under the design slamming pressure
f_{cLb}, f_{cb} and f_{cLT}	are critical buckling stresses with respect to uniform compression, ideal bending and shear, respectively, and may be determined in accordance with the following

The critical buckling stresses for rectangular plate elements, such as plate panels between stiffeners; web plates of longitudinals, girders, floors and transverses; flanges and face plates, may be obtained from the following equations, with respect to uniaxial compression, bending and edge shear, respectively.

$$f_{ci} = f_{Ei} \qquad\qquad \text{for } f_{Ei} \leq P_r f_{yi}$$
$$f_{ci} = f_{yi}\left[1 - P_r(1 - P_r)f_{yi}/f_{Ei}\right] \text{ for } f_{Ei} \leq P_r f_{yi}$$

where

f_{ci} critical buckling stress with respect to uniaxial compression, bending or edge shear, separately, in N/cm^2 (kgf/cm^2, lbf/in^2)

f_{Ei} $k_i\left[\frac{\pi^2 E}{12}\left(1 - v^2\right)\right]\left(\frac{t_n}{s}\right)^2$, in N/cm^2 (kgf/cm^2, lbf/in^2)

k_i buckling coefficient, as given in Table 9.4

E modulus of elasticity of the material, may be taken as 2.06×10^7 N/cm^2 (2.1×10^6 kgf/cm^2, 30×10^6 lbf/in^2) for steel

v Poisson's ratio, may be taken as 0.3 for steel

t_n net thickness of the plate, in cm (in)

s spacing of longitudinals/stiffeners, in cm (in)

P_r proportional linear elastic limit of the structure, may be taken as 0.6 for steel

$f_y\, f_y$ for uniaxial compression and bending

   $f_y/\sqrt{3}$ for edge shear

f_y specified minimum yield point of the material, in N/cm^2 (kgf/cm^2, lbf/in^2)

In the determination of f_{cL}, f_{cB} and f_{cLT}, the effects of openings are to be accounted for. A practical method of determining f_{cL}, f_{cb} and f_{cLT}, is the well-established eigenvalue analysis method with suitable edge constraints. If the predicted buckling stresses exceed the proportional linear elastic limit, which may be taken as $0.6 \times f_y$ for steel, plasticity correction is to be made.

Table 9.4 Buckling coefficient K_i

For critical buckling stress corresponding to f_L, f_T, f_b, or f_{LT}

I. Plate panel between stiffeners			K_i
A. Uniaxial compression 1. Long plate $\ell \geq s$		a. For $f'_L, = f_L$ b. For $f'_L, = \frac{f_L}{3}$ (See note)	$4C_1,$ $5.8C_1$
2. Wide plate $\ell \geq s$		a. For $f'_T, = f_T$ b. For $f'_T, = \frac{f_T}{3}$ (See note)	$\left[1 + \left(\frac{s}{\ell}\right)^2\right]^2 C_2$ $1.45\left[1 + \left(\frac{s}{\ell}\right)^2\right]^2 C_2$
B. Ideal bending 1. Long plate $\ell \geq s$			$24C_1,$
2. Wide plate $\ell \geq s$		a. For $1.0 \leq$ $\frac{\ell}{s} \leq 2.0$ b. For $2.0 < \frac{\ell}{s}$	$24\left(\frac{s}{\ell}\right)^2 C_2$ $12\left(\frac{s}{\ell}\right)^2 C_2$
C. Edge shear			K_i $\left[5.34 + 4\left(\frac{s}{\ell}\right)^2\right]C_1$

D. Values of C_1 and C_2
(1)For plate panels between angles or tee stiffeners
$C_1 = 1.1$
$C_2 = 1.3$ within the double bottom or double side*
$C_2 = 1.2$ elsewhere
(2)For plate panels between flat bars or bulb plates
$C_1 = 1.0$
$C_2 = 1.2$ within the double bottom or double side*
$C_2 = 1.1$ elsewhere

II. Web of Longitudinal or Stiffener Ki

(continued)

Table 9.4   (continued)

For critical buckling stress corresponding to $f_L, f_T, f_b,$ or f_{LT}	
A. Axial Compression	
Same as I.A.1 by replacing s with depth of the web and ℓ with unsupported span	
(a) For $f_L' = f_L$	$4C$
(b) For $f_L' = \frac{f_L}{2}$	$5.2C$
(see note)	
Where	
$C = 1.0$	for angle or tee stiffeners
$C = 0.33$	for bulb plates
$C = 0.11$	for flat bars

B. Ideal Bending	
Same as I.B.1 by replacing s with depth of the web and ℓ with unsupported span	$24C$
III. Flange and face plate	K_i
Axial compression	0.44

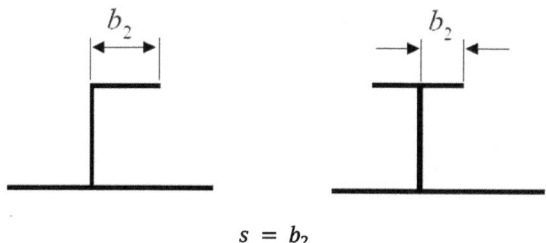

$$s = b_2$$

ℓ = unsupported span

* Applicable where shorter edges of a panel are supported by rigid structural members, such as bottom, inner bottom, side shell, inner skin bulkhead, double bottom floor/girder and double side web stringer

Note In I.A. (II.A), Ki for intermediate values of f_L' $/f_L$$(f_T'/f_T)$ may be obtained by interpolation between a and b

Correction to: Ship Vibration 4

Correction to:
F. Karkori, *Ship Vibration 4*, Synthesis Lectures on Ocean Systems Engineering,
https://doi.org/10.1007/978-3-031-74766-3

This book contains overlap in text with the previously published content [1] that was inadvertently omitted. The authors failed to attribute the reference [1]. The authors have now obtained permission to re-use this content from the American Bureau of Shipping.

Where [1] is: American Bureau of Shipping (2024), Rules and Guides https://ww2.eagle.org/en/rules-and-resources/rules-and-guides.html

The updated version of this book can be found at
https://doi.org/10.1007/978-3-031-74766-3

Annex: Documentation of Strength Assessment for Classification Review

A technical report is to be prepared to document the essential information used in the direct strength assessment and submitted to Classification for review. As a minimum, the documentation should typically include the following:

- A list of reference structural drawings, including dates and versions,
- Structural drawings,
  - General arrangement,
  - Midship section and typical transverse bulkhead,
  - Shell expansion,
  - Construction profile and deck plans,
  - Aft part key section,
  - Forward part key section,
- Lines or offset tables,
- Loading manuals,
- Loading conditions selected,
- Pressure points selected,
- Seakeeping programme used,
- Seakeeping input data including vessel speed, weight distribution, frequency range, wave headings, hydrostatic information, etc.,
- Seakeeping output data including motion RAOs for relative vertical motions and relative vertical velocities for the pressure points selected,
- Wave data and wave spectrum used,
- Maximum relative motions,
- Slamming pressure calculation programme,
- Design slamming pressures including time history plots,
- Results of direct strength assessment calculations. For main supporting members, the following finite element analysis information is also to be provided:

F. Karkori, *Ship Vibration 4*, Synthesis Lectures on Ocean Systems Engineering, https://doi.org/10.1007/978-3-031-74766-3

- The particulars of the finite element modelling, analysis and post-processing programmes used,
- Detailed description of finite element structural modelling and assumptions,
- Description of material properties,
- Description of load application and boundary constraints for hull girder and local sub load cases,
- Plots showing finite element meshing and scantlings,
- Stress/deformation plots of overall structural model and critical areas under applied loads to demonstrate the acceptance criteria are not exceeded,
- Results for buckling strength assessments of main supporting members,
- Component, von-Mises and principal stress plots of critical structural members/ details,
• Recommended modifications to the reference drawings and strength assessment results for modified structural members/details.

Classification may request detailed results and data files for verification and reference so that any suspected discrepancies can be quickly resolved.